Table of Contents

Picture Password:
A Visual Login Technique for Mobile Devices

Abstract: Adequate user authentication is a persistent problem, particularly with handheld devices such as Personal Digital Assistants (PDAs), which tend to be highly personal and at the fringes of an organization's influence. Yet, these devices are being used increasingly in corporate settings where they pose a security risk, not only by containing sensitive information, but also by providing the means to access such information over wireless network interfaces. User authentication is the first line of defense for a lost or stolen PDA. However, motivating users to enable simple PIN or password mechanisms and periodically update their authentication information is a constant struggle. This paper describes a general-purpose mechanism for authenticating a user to a PDA using a visual login technique called Picture Password. The underlying rationale is that image recall is an easy and natural way for users to authenticate, removing a serious barrier to compliance with organizational policy. Features of Picture Password include style dependent image selection, password reuse, and embedded salting, which overcome a number of problems with knowledge-based authentication for handheld devices. Though designed specifically for handheld devices, Picture Password is also suitable for notebooks, workstations, and other computational devices.

Introduction

The trend toward a highly mobile workforce has spurred the acquisition of handheld devices such as Personal Digital Assistants (PDAs) at an ever-increasing rate. These devices offer productivity tools in a compact form and are quickly becoming a necessity in today's business environment. Manufacturers produce handheld devices using a broad range of hardware and software. Handheld devices are characterized by small physical size, limited storage and processing power, and the means for synchronizing data with a more capable notebook or desktop computer.

Unlike desktops and notebook computers, handheld devices typically support a set of interfaces that are oriented toward user mobility. Usually they come equipped with a touch screen and a microphone, but lack a keyboard. One or more wireless interfaces, such as infrared or radio, (e.g., Bluetooth, WiFi, GSM/GPRS) are also built-in for wireless communication over limited distances to other devices and network access points. Most handheld devices can also send and receive electronic mail and browse the Internet via both wired and wireless interfaces. While such devices have their limitations, they are nonetheless extremely useful in managing appointments and contact information, reviewing documents, corresponding via electronic mail, delivering presentations, and accessing corporate data. Moreover, because of their relatively low cost, they are becoming ubiquitous within most organizations.

From an authentication perspective, several issues loom over the use of such device, including the following items:

- While handheld devices increasingly retain sensitive information over time and become networked to access wireless services and organizational intranets, some users may not recognize the associated security implications.

- Because of their small size, handheld devices may be left unattended temporarily, lost, or stolen.

- User authentication may not be enable, a common default mode, exposing the contents of the device to anyone who possesses it.

- Even if user authentication is enabled, the authentication mechanism may be weak (e.g., a four-digit PIN) or easily circumvented [Kin01].

- Once authentication is enabled, changing the authentication information for knowledge-based mechanisms (e.g., PIN) periodically as its lifetime expires is seldom done on the initiative of the user.

Adequate user authentication is the first line of defense for protecting the resources of a handheld device. Currently, many handheld devices come with a four-digit Personal Identification Number (PIN) and a numerical entry pad consisting of the digits 0-9 as a means for user authentication. Because of their limited length and alphabet, PINs may be susceptible to shoulder surfing or systematic trial-and-error attacks. Passwords offer a significant improvement over PINS in both length of entry string and size of alphabet. Some handheld devices, such as PDAs, support the character-by-character input of traditional alphanumeric passwords via the touch screen, using either handwriting recognition or a virtual keyboard window.

The strength of password mechanisms lies in the large set of combinations of character strings possible, from which an intruder has to identify the exact one needed to impersonate an authorized user. For example, for an eight-character string populated from the set of 95 printable ASCII keyboard characters, the number of possible character strings is 95^8. While passwords are an improvement over PINs, they can be difficult to remember and prone to input errors when entered via a touch screen. Thus, users tend to use easily remembered character strings such as "password" to simplify authentication. When the password expires, its replacement is often very similar (e.g., p@ssword) [Lee01]. These tendencies significantly reduce the password space and allow an intruder to deduce passwords quickly by systematically applying dictionaries of commonly used strings and password reuse patterns (e.g., password1, password2, etc.) [Mor79, Kle90].

To avoid weak or easily broken passwords, organizational policy and procedures compel users to include special, upper case, and numeric characters in their password string, to avoid common or easily guessed strings, and to update passwords regularly (e.g., every 90 days) with completely different strings. Policy and procedures may also be backed up by technical controls that force periodic updates, and either screen out unacceptable passwords selected by users or supply acceptable passwords automatically for users [Spa92]. Unfortunately, the measures put in place to ensure strong, but typically complex and meaningless passwords, frequently result in users writing them down and keeping them near the computer system to recall quickly. This behavior is fueled further by the numerous systems that require passwords for authentication.

The mechanism described in this paper authenticates a user to a PDA using a visual login technique called Picture Password. Picture Password authenticates a user through the selection of images displayed on a handheld device. As seen with screen savers, graphic screen backgrounds, or application skins, products that employ visual content appeal to a large class of users. Having the ability to tailor the display interface with personal images, gives users a sense of freedom and control over what might otherwise be considered an imposition. Thus, an image-based authentication technique, if designed carefully, has the potential to engage users to employ the mechanism and indulge periodic updates of the authentication information.

Overview

For any authentication mechanism to gain user acceptance, it must be convenient to use and match the capabilities of the device. Difficulties due to cumbersome attachments, slow performance, or error-prone user interfaces are typically not tolerated. The aim then is to devise authentication mechanisms for PDAs that are well suited to the typical interfaces and capabilities supported by such devices. Our work has focused on higher-end expandable devices aimed at corporate users: the Compaq/HP iPAQ H series of devices running the Familiar Linux distribution and Opie and the Sharp Zaurus SL-5X00 series of devices with Lineo's embedded Linux and Qtopia. Both families of devices are approximately the same dimensions as a pocket-size agenda, equipped with a one-quarter VGA touch screen, use processors running at 200 MHz and higher, and have comparable amounts of read only flash memory (32MB or more) and random access memory (64MB or more). Linux distributions are supported on a number of other PDAs including the CDL Paron, IBM e-LAP, and Yopy Gmate.

The Picture Password authentication mechanism has two distinct parts: the initial password enrollment and subsequent password verification. During enrollment, a user selects a theme identifying the thumbnail photos to be applied and then registers a sequence of thumbnail images that are used to derive the associated password. When the device is powered on or booted up, the user must enter the currently enrolled image sequence for verification to gain access to the device. After a successful authentication, the user may change the password, selecting a new sequence and/or theme.

Picture Password offers benefits over PINs and textual passwords, especially for the visually inclined user. As with textual passwords, a similar password length and alphabet size is used. However, instead of having to memorize and enter a string of random-like alphanumeric characters, a sequence of thumbnail images must be selected and retained. Experimental results suggest that human visual memory is well suited to such visual and cognitive tasks [Mel01, Gol71]. Moreover, an image sequence that has some meaning to the individual user (e.g., logos of sport's league teams in order of preference, one's family members in order of birth or vacation spots in order visited) can be used. If forgotten, the sequence may be reconstructed from the inherent visual cues. The display interface presents images in an easy-to-select size, reducing error entries. The underlying mechanism, which handles random character code assignment to images, password composition, enrollment, and verification, is completely hidden from the user.

The presentation of images to the user for selection is based on tiling a portion of the user's graphical interface window. Various ways exist to tile a surface with both regular and irregular patterns. Picture Password uses squares of identical size (40 x 40 pixels), grouped into a 5 x 6 matrix of elements. Cells of this size provide a clear recognizable image that is easily selectable for most users. Figure 1 illustrates the PDA screen image for the Cats & Dogs theme, one of three predefined themes. The message area at the top of the window guides the user's actions, while the display area in the center displays thumbnail images selectable with a single tap of the stylus. The controls at the bottom allow the user to clear out any incorrect input entered or submit the entered image sequence for verification. Selecting and submitting the correct sequence of thumbnail images authenticates the user to the device.

Figure 1: Picture Password Cats and Dogs Theme

Picture Password allows users flexibility in choosing a predefined theme that suits their personality and taste or providing their own set of images for display. All thumbnail images must be in a predefined digital format, which can be created using an image manipulation tool such as Photoshop or GIMP. Besides a random layout of individual thumbnail images, several thumbnail images may be structured to compose a single composite image as in a mosaic, where each thumbnail image contributes a portion to some larger image. Figure 2 illustrates the Sea & Shore theme, where all 30 thumbnail images in the display area form a single contiguous image.

Figure 2: Sea & Shore Theme

Image Selection

A perplexing problem faced in Picture Password was ensuring that its password space would be comparable to that of alphanumeric passwords. The size of the image matrix limits the effective alphabet size to only 30 elements, assuming a one-to-one mapping, which in turn results in weak passwords. For example, an eight-entry image sequence, the number of possible password strings would be only 30^8. Therefore, several ways to increase the alphabet size were considered. They included allowing passwords to be composed with thumbnail elements from all of the three available themes (90 elements total) and providing more images per theme, by either using smaller-sized images, or adding a feature to zoom up larger composite images from each thumbnail image. Each of these approaches had serious drawbacks. Using thumbnail images from all three themes would require more difficult navigation for the user. A denser set of images would mean less tolerance for selections and a higher rate of input errors. Zooming up a thumbnail image to a larger composite image would require more user interaction when selecting images and greater complexity in creating and handling themes.

The main criterion for selecting among various alternatives was to maintain the simplicity of the user interface, keeping it as easy to use as possible. Our solution was to add a second method or style for choosing thumbnail elements. Besides selecting individual thumbnail elements as before, one could now select two thumbnail elements together to compose a new alphabet element. The concept is akin to using a shift key to select uppercase or special characters on a traditional keyboard, but in this context each thumbnail element serves as a shift key for every other element, including itself. With this addition, the resulting alphabet size expands from 30 elements (i.e., singly selected keys) to 930 elements (i.e., 30 singly selected keys plus 30 x 30 composite keys), which compares favorably to the 95 printable ASCII characters available from a traditional keyboard. Several ways exist to select a pair of buttons and link them in composing an alphabet element. Drag-and-drop is perhaps the most obvious method, but not typically supported by all handheld device windowing systems. Another, more generic selection method is choosing the first thumbnail image by picking and holding the stylus there, highlighting the selection, and then completing the pair by picking the second button image in the normal fashion.

Having someone observe the user entering a password and using that information to gain entry is a concern with any password system. Fortunately, the screens of PDAs and most handheld devices have a narrow viewing angle. That property, combined with their small size, makes it relatively easy to shield data entry with one's body. Nevertheless, observation is a concern. As a safeguard, Picture Password gives users the option to have images shuffled automatically between authentication attempts, where appropriate. Supporting two different styles of selection also is a safeguard, since it makes it difficult for an observer to glean both the entire image sequence and the selection style for each image in the sequence.

Password Formation and Reuse

Organizational policies typically require users' passwords to expire and be changed completely after some period of use. This practice keeps a persistent intruder from cracking a password over some indefinite lifetime of use. Though effective, password expiration is also a nuisance for the user, who follows this practice on numerous systems and accounts, and continually must forget old and memorize new passwords. The user would instead prefer to continue using the same image sequence indefinitely.

One solution for password reuse is to allow the same image sequence to be used after it expires, but have the image sequence generate a completely new password value. By decoupling images from alphabet, numerous distinct mappings between those respective sets of elements are possible. To enable password reuse, the authentication mechanism has only to select a mapping that results in a different password value to be generated for the same image sequence.

As shown in Figure 3, during initial enrollment or a subsequent reenrollment, for each thumbnail element of the image matrix, Picture Password randomly assigns a distinct value from the full range of possible alphabet values to form a value matrix with the same dimensions. While the set of image elements is fixed at 30, the set of basic alphabet values from which the 30 needed are drawn can be significantly larger. Thus, the elements of the value matrix contain the basic alphabet used to compute the password, but are independent from the image matrix.

The mapping of thumbnail element to value element remains constant from one authentication attempt to another and changed only during reenrollment. Because values are randomly associated with each thumbnail element of the image matrix during reenrollment, selecting the same theme and sequence of thumbnails repeatedly should produce a completely different password value. As an added measure, mappings that produce the same password value as the one previously enrolled can be rejected during reenrollment.

Image Matrix

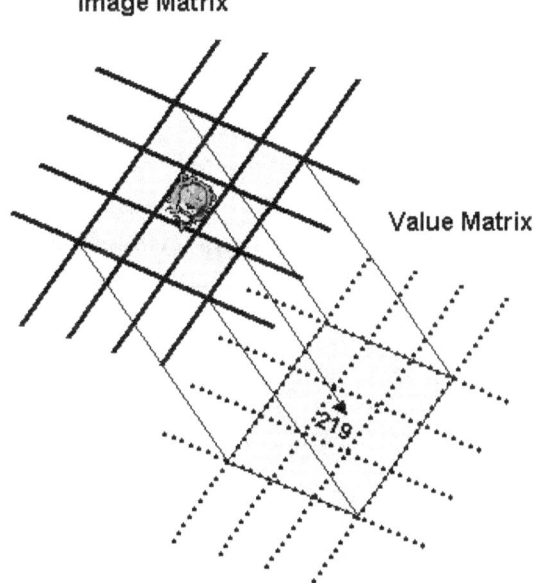

Value Matrix

Figure 3: Image and Value Matrices

The sequence of thumbnail elements selected by the user in either an enrollment or authentication attempt governs password formation. The mapped value of a single image selection can be directly applied, while the two mapped values of a paired image selection must first be composed into a single value. For example, if single-byte, non-zero, unsigned integers comprise the set of 30 basic alphabet values, a single image selection could be the pair of bytes $(0, VM_j)$, and a paired image selection could be the pair of bytes (VM_k, VM_j), where VM_j represents either the alphabet value in the value matrix for a single image selected or the first image selected in a paired image selection, and VM_k represents the alphabet value for the second image in a paired image selection.

Once the thumbnail images for an image sequence have their alphabet values resolved, those values are concatenated together, in the sequence the images were selected, to form the clear text password. Picture Password then applies a one-way cryptographic hash to the resulting string iteratively to form the password. The NIST Secure Hash Algorithm (SHA) is used to compute the cryptographic hash and results in a 20-byte binary value. The number of iterations to apply the hash algorithm is controlled by a variable to allow the work effort to be adjusted to the level of security needed. The user's password is never maintained in unencrypted form on the device. Instead, only the hash result is retained during enrollment and used later during verification to compare against the hash result from any subsequent authentication attempt.

While a visual login technique by its very nature avoids dictionary attacks associated with textual passwords, it may be possible for an intruder to compile commonly used set of image selections (e.g. location-based sequences such as the four corners or main diagonal of the image matrix) and use them in an attack. As a countermeasure to an intruder directly applying a dictionary of commonly used passwords, the clear text password value may be prepended with a random value, commonly referred to as a salt, before the hash is iteratively applied. This step significantly increases the work factor for the intruder, in relationship to the size of the salt value that is used and whether both a public and a secret salt are involved [Man96, Aba97].

One additional use for the value matrix is to hold individual salt values for each element of the basic alphabet. Rather than prepending the resulting clear text value of the password with a collective salt value, salting can be done continuously as entries in the value matrix are used to generate a password value. This method of salting takes advantage of any unused memory within the value matrix, in situations where the memory allocated for each value matrix element is sufficient to hold both an alphabet value and its associated salt value. Otherwise, additional memory can be allocated to hold the salt component, which can be treated simply as the prefix of an alphabet value. Each time a new password is enrolled, the salt components are populated with random values. This procedure, in effect, creates a new way of salting the password through the embedding of salt values within the alphabet value entries of the value matrix.

Password Strength

As mentioned earlier, with 30 thumbnail images to choose, the effective size of the alphabet is 930, (30 + (30*30)). Passwords formed with so large an alphabet space are quite strong. Thus, 7-entry long passwords have 930^7 possible values for a password space of approximately 6.017009e+20, which is an order of magnitude greater than that for 10-character long, alphanumeric passwords formed from the 95 printable ASCII character set, which is 95^{10} or approximately 5.987369e+19. The general strength relationship between visual passwords formed from a 30-element matrix versus textual passwords formed from the 95 printable ASCII characters is approximately

$$N_{pp} = \lceil {}^2/_3 * N_{tp} \rceil,$$

where N_{tp} is the required character length for textual password input, N_{pp} is the corresponding input sequence length required for Picture Password, and $\lceil x \rceil$ is the "ceiling" function. In simple terms this means that image sequences formed with dual selection styles require approximately one-third less length than that of a traditional alphanumeric password. This presumes, of course, that just as additional keystrokes are needed to select special and capital characters on a keyboard, a comparable number of additional strokes are used when forming an image sequence involving paired image selections. Table 1 gives a comparison of input lengths between the two mechanisms for a range of sizes of password elements. Note that the values in the table presume that just as additional keystrokes are needed to select special and capital characters on a virtual keyboard, a comparable number of additional strokes are used when forming an image sequence involving paired image selections.

Table 1: Input Length Comparison

Textual Password Length	6	7	8	9	10	11	12
Image Sequence Length	4	5	6	6	7	7	8

8

Mechanism Protection

Picture Password relies primarily on two forms of authentication information: the cryptographic hash of the password string computed from the enrolled image selection, and the value matrix that maps selected thumbnails to their underlying alphabet values, each containing an embedded salt. The confidentiality and integrity of both pieces of information must be protected by the underlying operating system. At a minimum, this information is safeguarded through strict file access control settings. However, should read access to this information be gained somehow, the authenticating image sequence is inherently resistant to discovery, requiring a difficult and exhaustive brute force effort to uncover.

Access controls alone may not provide sufficient protection for all handheld device environments. Other, more subtle, ways exist that an intruder might use to foil the authentication mechanism. The following items list some common threats that should be safeguarded against to minimize the possibility of a successful attack on an implementation of Picture Password:

- Personnel Information Management (PIM) or other applications that run as root by default may be exploited as a potential avenue for gaining access to the authentication information, as well as other information.

- Program binaries may become compromised and replaced with Trojan versions that capture user input or intermediate clear text forms of the password.

- Program interfaces, system resources, and device interfaces may be manipulated to cause the mechanism to be bypassed completely or to fail in a way such that it does not perform the password computation procedure correctly and is compromised.

- Spoofing may be used to grab user input either when images are being selected or when the image sequence is being handled.

- Savaging or sniffing may be used to obtain user input either when images are being selected, when the image sequence is being handled, or after the authentication mechanism terminates execution.

Implementation

Picture Password was implemented in C++ for a Linux iPAQ PDA, and for the Open Palmtop Integrated Environment (Opie), an open source implementation of the Qtopia graphical environment of TrollTech. The design of the authentication solution involves three main types of components: kernel modifications, authentication handlers, and user interface (UI) components. Figure 4 illustrates the different parts of the solution and the flow of data between them. Note that the design supports multiple authentication mechanisms running in series, whereby each mechanism comprises a matching pair of user interface component and authentication handler. The discussion in this section, however, focuses solely on operating a single authentication mechanism, Picture Password.

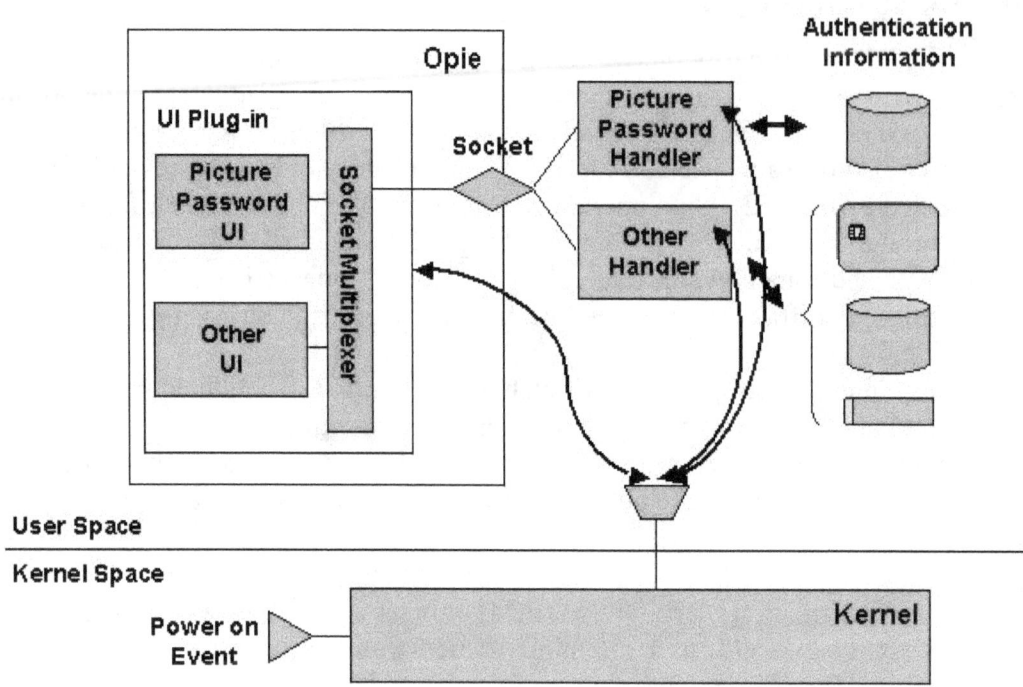

Figure 4: Design Overview

The responsibility for determining when authentication is required, by monitoring sleep/wake-up events and signals from the Opie plug-in module, falls to the kernel. The Linux kernel was patched to initiate user authentication through a set of registered authentication handlers each time the device is rebooted or powered on, starting and suspending each handler in sequence as needed. The kernel was also modified to block the I/O ports on the device and lock down other means to bypass the authentication process until the user successfully completes authentication. One byproduct of these modifications is that the power key also takes on the role of a trusted path mechanism for asserting the authentication mechanism user interface.

The kernel patches needed to support device lockdown were developed earlier as part of a general scheme to enforce corporate policies on handheld devices [Jan03]. Policy controls restrict access to authentication information to the appropriate handler and also prevent the code for the plug-in, user interface components and handlers from being deleted or replaced. The policy controls apply equally to both root and non-root processes, and also encompasses all communications and peripheral attachment interfaces. Another kernel enhancement periodically checks whether the authentication handlers are running, and restarts them if they were terminated, which makes the implementation quite robust and able to accommodate failure conditions that might otherwise cause the authentication mechanism to malfunction or be bypassed.

The user interface for an authentication mechanism is implemented as a component within a plug-in module developed specifically for Opie. As the name implies, the function of the user interface component is to interact with the user under the control of its associated handler. For example, in Picture Password, the user interface displays the image matrix and obtains the image sequence entered by the user, which it returns in a response to the handler. The plug-in module supports a socket interface to receive commands from the authentication handler

10

components that run as separate processes, and route the commands to the correct user interface component within the plug-in. Similarly, the reverse response process is also supported between components and the module. Communication between the plug-in module and the various user interface components is done using Qte's signal and slot facility. The user interface module, as a plug-in to the desktop environment, is loaded automatically by Opie upon system boot up and share its address space.

Handlers embody the core mechanism that performs the actual authentication. They interact with the user interface components to tell them to bring up the specific screens, accept input, display messages, etc. Handlers also have responsibility for interactions with tokens, smart cards, the file system, etc., needed to perform the authentication. In the case of Picture Password, the handler has exclusive access to a file containing the theme identifier, value matrix, and the password hash, which it uses to enroll a user's password and to verify authentication attempts. Handlers communicate with the kernel module, listening when to initiate authentication, and reporting if the authentication was successful.

The process startup and synchronization among components proceeds as follows:

- On system boot-up, the kernel loads and enforces its default policy, which blocks I/O ports on the device, hardware keys, and access to the authentication handler's code, as well as restricts access to authentication information within the file system exclusively to the appropriate authentication handler. The Linux proc file system (/proc) provides a communication channel between user space processes (UI components and handlers) and the kernel module. The kernel registers a file in the /proc file system for user space processes to trigger actions in the module.

- The system startup script tells the kernel (through the /proc/policy file) the filenames of the handler and any other related programs that need to be active. The kernel sees that the handler programs are not running and starts them.

- Upon startup, each handler program performs all necessary initialization and then reads from the /proc file entry, which causes their execution to be suspended.

- Opie and its plug-ins are also loaded during boot-up. The UI plug-in module signals the kernel via the /proc file entry that the device should be locked and retrieves the list of registered handlers with which it can communicate.

- At this point all the components of the system are running and the default set of least privileges are being enforced.

- The kernel wakes up the first authentication handler, Picture Password, to resume processing.

- The Picture Password handler reads the authentication information from the file system and signals its user interface component via a socket interface with the identity of the theme to display.

11

- The user interface component displays the theme, accepts the image sequence, and returns that information to the handler.

- The handler uses the image to alphabet mapping to compute and verify the password. If the authentication attempt is successful, it reports success to the kernel module via the /proc interface and removes the authentication window from the screen. If unsuccessful, it continues to use the user interface component to have the user retry until a successful authentication is completed.

- When the kernel receives an indication of success from the handler, it suspends it, and initiates any other registered handlers. If this is the last handler, the kernel unlocks the device.

Picture Password has also been ported to the Pocket PC 2002 operating environment. For this implementation, existing procedures and interfaces for substituting a custom password mechanism for the default alphanumeric password mechanism could be followed [Mic03]. The main differences from the Opie/Linux implementation are that a single Windows applet module embodies both the user interface and handler portions of Picture Password; the programming interface takes care of storing and checking password values and asserting the mechanism at power on; and some device registry settings are used in addition to the information kept within the file system.

Related Work

United States Patent 5,559,961 [Blo96] describes a system and method for applying graphical passwords. Rather than using a distinct matrix of thumbnail images as in Picture Password, the mechanism displays the image areas or cells that make up a single graphical image. The user selects predetermined areas of an image in a correct sequence, as a means of entering a password. The password is composed during enrollment by allowing the user to position selected cells from the image in a location and sequence within the display interface. The mechanism stores the sequence of cells from the image as a password. The cells are removed from the display when enrollment is completed, leaving only the original image. No discussion is given of how the image cells are used to form a password value or the security of the purposed scheme. One drawback appears to be that the cells, which in effect form the alphabet for composing a password, offer a significantly smaller sized alphabet than that available for both alphanumeric passwords and Picture Password.

Draw-a-Secret (DAS) [Jer99] is a scheme for graphical password selection and input, targeted for PDAs. The user draws a design on a display grid from the interior of one cell to another, which is used as the password. Each continuous stroke is represented as the sequence of cell grids encountered. Strokes can start anywhere and go in any direction, but must occur in the same sequence as the one enrolled for the user. The size of each cell must be sufficiently large to allow the user a degree of tolerance when drawing a graphical password. Each continuous stroke is mapped to a sequence of coordinate pairs by listing the cells through which it passes, in the order in which the stroke traverses the cell boundary. The grid sequences for each stroke that composes a drawing are concatenated together in the order they were drawn to form a password. The size of the password space for graphical passwords formed using this scheme was shown to be, generally speaking, larger than that of textual passwords. Let-Me-In, an example password replacement mechanism from Microsoft, works in a similar fashion to DAS, using a grid of points rather than distinct cells, whereby the user enters a pattern of connecting grid points to form a password [Mic03]. While the example was done mainly to illustrate how to implement a replacement mechanism within the Pocket PC environment, it also suggests an alternative user interface that could be used for the DAS framework.

Déjà Vu, a project at the University of California Berkeley, also involves using a set of images for user authentication [Dha00]. Rather than using real-life images, abstract images are generated randomly using a hash visualization technique [Per99]. During enrollment, the user selects a set of images that makes up his authentication base. A training phase is then used to improve a user's recognition of the abstract images within his authentication base. The authentication mechanism is an n-out-of-m recognition scheme, whereby the user must identify a selection of the images from the authentication base when presented to him within a much larger challenge set containing decoy images. A trusted server stores the authentication base for each user and provides the challenge set for each attempted user authentication, which make this scheme unsuitable for handheld devices, which may have only intermittent network connectivity. The server must be tightly secured to guard the confidentiality of the authentication information or the scheme fails entirely. To counter shoulder surfing, different sets of images, both legitimate and decoy, may appear in random positions of the display for each authentication attempt. Passface, a commercial user authentication system from Real User, is a challenge-response mechanism somewhat similar to Déjà Vu, but uses the faces of individuals instead of abstract images [Rea01].

A commercial product called visual Key [sfr00], from sfr GmbH in Cologne Germany, uses cells of a single predefined image as the password elements. The visual Key software forms a selection matrix by dividing a single image into cells and dynamically adjusting the grid so that cell centers align with the touch point during selection. A user must select a specific sequence of cells from the display to be granted access to the device. The strength of the password depends on the number of cells that make up the image, since they are used to determine the range of the password alphabet. Approximately 85 distinct cells with a size of 30 x 30 pixels can fit on a standard size 240 x 320 pixel display of a PDA, which results in an alphabet size smaller than the 95 printable ASCII characters available with both alphanumeric passwords and Picture Password. Cells comprised of 30 x 30 pixels or less are a bit small, which can contribute to selection errors. Another drawback is that during selection the cells are not made visible, requiring user to remember which part of an object in the image to select (e.g., the upper left corner of a door or window), if the object encompasses more than one cell. In contrast, Picture Password uses distinct buttons 40 x 40 pixels in size, instead of cells, so that user selections are clearly registered, making it more likely that the user enters the password quickly and correctly on the first attempt.

PointSec for Pocket PC is a commercial product that includes several authentication-related components that can be managed centrally [Poi02]. PicturePIN is a graphical counterpart to a numeric PIN system that uses pictograms rather than numerics, for entering the PIN via a keypad-like layout of 10 keys. The symbols, which can be tailored, are intended to form a mnemonic phrase, such as the four-symbol sequence of ⚲ - men / ❤ - love / ☝ - to listen / ♫ -

to music. The sequence of symbols can be between 4 and 13 symbols long, and to increase security against "shoulder surfing," the symbols are scrambled at each login. As an added usability feature, QuickPIN enables fast access to mobile devices within a specified number of minutes, between 30 and 300 seconds, after last power off. QuickPIN relies on a minimum of two pictogram symbols to allow users access to their PDA. Both the PicturePIN and QuickPIN systems can be set to lock out a user after three to an infinite number of unsuccessful attempts. PicturePIN is somewhat similar to Picture Password, but supports only a limited alphabet size and a single selection style, making it a far less powerful mechanism in and of itself. Moreover, Picture Password can be configured with themes composed of pictogram symbols to support mnemonic password phrases. SafeGuard PDA is another commercial product whose Symbol PIN authentication option works very similarly to PicturePIN and with similar limitations [Uti03].

Summary

Picture Password is a visual login technique that matches the capabilities and limitations of most handheld devices and provides a simple and intuitive way for users to authenticate. Besides user authentication, Picture Password may also be used in other security applications where conventional passwords have been used traditionally. For example, password-based encryption, whereby a password value is transformed into a cryptographic key suitable for encrypting files or other information could rely on keys derived by the image selection technique of Picture Password. The mechanism is relatively straightforward and flexible, and one that is suitable for many users, organizations, and types of devices. The approach provides a simple and hopefully entertaining way for users to authenticate to a device, which should remove much of the burden associated with employing an authentication mechanism. Moreover, with style dependent image selection, password reuse, and embedded salting, the mechanism is superior to using traditional ASCII passwords of comparable length. While the solution is particularly well suited for handheld devices, Picture Password can also be used in a wide range of platforms, from appliances having single embedded processors, to large-scale multi-processor computers.

References

[Aba97]   Martın Abadi, T. Mark A. Lomas, and Roger Needham, Strengthening Passwords, SRC Technical Note 1997 – 033, digital Systems Research Center, December 1997.

[Blo96]   Greg E. Blonder; Graphical password, US Patent 5559961, Lucent Technologies Inc., Murray Hill, NJ, August 30, 1995.

[Dha00]   Rachna Dhamija and Adrian Perrig, Déjà Vu: A User Study Using Images for Authentication, Proceedings of the 9th Usenix Security Symposium, August 2000.

[Gol71]   Alvin Goldstein and June Chance, Visual Cognition for Complex Configurations, Perception and Psychophysics, 9, 1971, pp. 237-241.

[Jan03]   Wayne Jansen, Tom Karygiannis, Michaela Iorga, Serban Gavrila, and Vlad Korolev, Security Policy Management for Handheld Devices, The 2003 International Conference on Security and Management (SAM'03), June 2003.

[Jer99]   Ian Jermyn, Alain May, Fabian Monrose, Michael Riter, and Avi Rubin, The Design and Analysis of Graphical Passwords, Proceedings of the 8th USENIX Security Symposium, August 1999.

[Kin01]   Kingpin and Mudge, Security Analysis of the Palm Operating System and its Weaknesses Against Malicious Code Threats, USENIX Security Symposium, August 2001.

[Kle90]   Daniel Klein, Foiling the Cracker: A Survey of, and Improvements to, Password Security, 2nd USENIX Unix Security Workshop, August 1990, pp. 5-14.

[Lee01]   Jennifer 8. Lee, And the Password Is . . . Waterloo, The New York Times, Circuits, Thursday, December 27, 2001, Late Edition - Final, Section G, Page 1, Column 1, <URL: http://www.nytimes.com/2001/12/27/technology/circuits/27PASS.html?ex=10281 05054&ei=1&en=4c3ce4a63682fecc>.

[Man96]   Udi Manber, A Simple Scheme to Make Passwords Based on One-Way Functions Much Harder to Crack, Computers & Security, 15(2), 1996, pp. 171-176.

[Mic03]   Let Me In: Pocket PC User Interface Password Redirect Sample, Microsoft Knowledge Base Article – 314989, Microsoft Corporation, July 2003, <URL: http://support.microsoft.com/default.aspx?scid=kb;en-us;314989>.

[Mel01]   David Melcher, The persistence of visual memory for scenes, Nature, 412(6845), p. 401(July 2001).

[Mor79]   Robert Morris and Ken Thompson, Password Security: A Case History, Communications of the ACM, Vol. 22, No. 11, November 1979, pp. 594-597.

[Per99]   Adrian Perrig and Dawn Song, Hash Visualization: a way to improve real world security, International Workshop on Cryptographic Techniques and E-Commerce, CrypTEC '99, 1999.

[Poi02]   Pointsec for Pocket PC, Pointsec Mobile Technologies, November 2002, <URL: http://www.pointsec.com/news/download/Pointsec_PPC_POP_Nov_02.pdf>.

[Rea01]   The Science Behind Passfaces, Document Revision 2, Real User Corporation, September 2001, <URL: http://www.realuser.com/published/ScienceBehindPassfaces.pdf>.

[sfr00]   visual Key – Technology, sfr GmbH, 2000, <URL: http://www.viskey.com/technik.html>.

[Spa92]   Eugene Spafford, OPUS: Preventing Weak Password Choices, Computers & Security, Vol. 11, No. 3, May 1992, pp. 273-278.

[Uti03]   SafeGuard PDA, Utimaco Safeware AG, March 2003, <URL: http://www.utimaco.com/eng/content_pdf/sg_pda_eng.pdf>.